Jellyfish and Other Sea Creatures

by Oxford Scientific Films

Photographs by Peter Parks

G. P. Putnam's Sons   New York

First American edition 1982
Text copyright © 1981 Jennifer Coldrey
Photographs copyright © 1981 Oxford Scientific Films Ltd.
Printed in Belgium by Henri Proost & CIE PUBA
Library of Congress Cataloging in Publication Data
Jellyfish and other sea jellies.
Jellyfish and other sea creatures.
Previously published as: Jellyfish and other sea jellies. 1981.
Summary: Examines, in text and photographs,
the physical characteristics and life cycles
of the jellyfishes and related sea creatures.
1. Medusae—Juvenile literature.   2. Coelenterata—Juvenile literature.
[1. Jellyfishes.   2. Coelenterates]   I. Parks, Peter, 1942–ill.
II. Oxford Scientific Films.   III. Title.
QL377.S4J44   1982   593.7   81-10672 AACR2
ISBN 0-399-20852-6
First impression.

Grateful acknowledgment is made to J. A. L. Cooke for
photographs reproduced on pages 26 (left) and 28, and to
Laurence Gould for those on pages 27, 29 (right) and
32 (bottom).

Jellyfish and Other Sea Creatures

Jellyfish and other sea jellies are a large and complex group of animals. They range in form from the colorful anemones and corals, most of which spend their life attached to the seabed, through many types of tiny free-floating creatures to the larger and more common types of jellyfish we sometimes find stranded on the beach. They are mostly marine, although one or two fresh-water species are known.

They are all coelenterates, a word which literally means "hollow gut." Although the shape may vary, each animal is essentially like a flexible gelatinous, hollow bag with a ring of tentacles around the mouth. The tentacles capture food and pass it into the bag, or stomach, for digestion. Any undigested remains are passed out through the same opening by muscular contraction of the body wall. The wall is basically made of two layers of cells — an outer protective "cover" and an inner lining to the stomach. This inner lining secretes digestive juices and also absorbs food. Between these two layers of cells, and secreted by them, is a jellylike material (called the mesogloea) which is sometimes very thin, or, as in most of the big jellyfish, is very thick, forming the bulk of the animal.

The outer and inner layers of cells contain muscle fibers which enable the animal to move and to expand and contract its body. There is also a simple nerve network in these layers.

All coelenterates possess stinging cells called nematocysts, which are found mostly on the tentacles and which are used for capturing prey as well as for defence against enemies. Each nematocyst is like a miniature harpoon loaded with poison, which is shot out at any creature that is unfortunate enough to touch the little trigger hair on the outside of the cell. Before it emerges, the long, hollow stinging thread is coiled up inside the cell. There are several sharp barbs at the base of this thread, and as the nematocyst emerges, these barbs latch onto the victim while the poisonous stinging thread is shot into its body like a hypodermic needle.

Large numbers of these stinging cells are often packed closely together on the tentacles, and there are other types of nematocyst, too. Some wrap around and entangle the prey, while others are sticky enough to hold onto it. These types have longer trigger hairs and first capture the prey after which the stinging nematocysts deliver their paralyzing blow. If the victim is small, it is quickly overcome and carried to the mouth by the tentacles. If it is larger it may manage to escape with a painful reminder of its dangerous encounter.

All coelenterates are carnivorous, trapping and eating small animals such as microscopic plankton, larval fish, and even other small jellyfish.

One of the simplest forms is the polyp, easily recognized as the sea anemone and the tiny freshwater animal, the hydra.

Most polyps reproduce simply by budding. A new individual is formed which eventually breaks off and drifts away to settle and grow into an adult. In many marine polyps the new individuals stay attached to their parent, so that little branching colonies are built up. Corals are like this, and there are many other marine polyp colonies which can be found growing on seaweeds, rocks and piers along the coast as well as on the bottom of shallow seas. Some look like mosses or lichens covering the rocks, while others have the appearance of miniature ferns. Some individuals within a polyp colony give rise to a different form, called a medusa.

The medusae are usually set free to float and drift among the plankton of the sea, where they produce eggs and sperm. The fertilized egg will eventually develop into a tiny sausage-shaped larva covered with fine hairs. This planula larva may be carried tremendous distances in the ocean currents before finally coming to rest on the bottom. Once settled, it attaches itself to the substrate at one end, while at the opposite end a mouth and tentacles appear and it develops into a polyp. This complicated life history, which is typical of the true jellyfish, is most easily explained in a diagram. The polyp and medusa stages are two different forms of individual, each living a totally different sort of life.

In some coelenterates the polyp and medusa stage are both important in the life cycle, but there are many others in which only one stage seems to be present. Thus, the sea-anemones and corals have no medusoid stage and exist only in polyp form, while in the true jellyfish, the medusa is the dominant stage in the life cycle.

There are many other complicated variations in the life cycle of these animals — so many, in fact, that scientists still do not know the full life history of many species; and because the medusa and polyp stages are often found so far apart, it is extremely difficult to decide whether or not they are two different forms of the same animal.

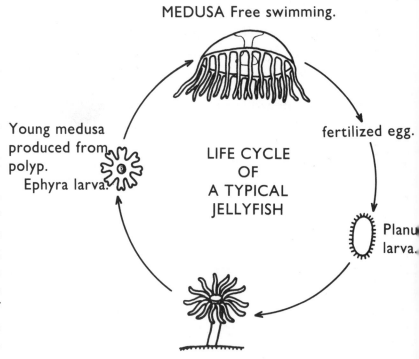

MEDUSA Free swimming.

fertilized egg.

LIFE CYCLE
OF
A TYPICAL
JELLYFISH

Young medusa produced from polyp.
Ephyra larva.

Planula larva.

POLYP Attached to bottom.

The medusa is what we commonly think of as a "jellyfish" and is so named because it resembles the mortal lady in Greek mythology whose hair was turned into snakes by a jealous goddess. It is a different shape from the polyp, having a much wider, flatter and more circular body, rather like a mushroom. The mouth hangs down on a short stalk (manubrium) from the center of the body, and this opens into the stomach, which, instead of forming one central cavity, is rather compressed in the center, but extends along radial canals to the margins of the mushroom, where a circular food canal runs around the edge. This canal system carries food from the central stomach to all parts of the animal.

Compared to the polyp, the tentacles on the outer rim are widely separated from the mouth, and in some medusae another set of large and often frilly tentacles grows from the manubrium.

The jelly layer in the medusa is much thicker than in the polyp, especially on the upper side, and this gives the body a soft and rubbery rigidity which helps the floating medusa to withstand the buffeting of the sea. Although these jellyfish spend much of their time floating like parachutes at the mercy of ocean currents, they can also swim actively and have a well-developed muscular system which brings about rhythmic pulsations of the body, propelling the animal through the water.

Many medusae have sense organs around the margin of the "mushroom" which are connected to the main nerve network. At the base of the tentacles patches of darker cells, which are light-sensitive, can often be found, and associated with these, or at other points around the rim, are special organs of balance, which tell the animal which way is up and enable it to correct its position if it is tilting in the wrong direction. These balance organs (statocysts) can also detect vibrations in the water and may enable the medusa to locate the position of small prey swimming nearby. The radial arrangement of the body means that the medusa can make use of food coming from any direction.

The sex organs of male and female medusae are borne either on the radial canals or, more commonly, on the manubrium. The sperm from a passing male medusa are attracted, probably by some chemical, into the female ovaries, and here the eggs are fertilized. They are commonly brooded for a while on the frills of the oral tentacles before being released into the sea as short-lived planula larvae. The planula settles and grows into a little polyp which eventually buds off tiny free-floating medusae called ephyra larvae. Swarms of these are often found in the plankton. When young they have no proper tentacles and feed by sucking prey directly into the mouth.

There is a great variety in the sizes, shapes and colors of jellyfish medusae. Some are less than half an inch in diameter, while others may be as large as six feet across. Some are a flattened umbrella shape, while others are a deep bell shape, often with a wide rim projecting inward around the margin, which greatly restricts the opening to the bell. These jellyfish are capable of quite rapid swimming, forcing water out of the constricted bell opening and moving by jet propulsion through the sea.

The number and size of the tentacles is also very variable. For example, tentacles up to a hundred feet long have been measured on that dangerous creature the Portuguese man-of-war. The Portuguese man-of-war (*Physalia*) is not a simple medusa, but a complicated collection of medusoid and polyplike "persons" which act together to form the whole animal. This colonial creature floats at the surface of the sea by means of an air-filled bladder with a diagonal crest along the top. This serves as a sail and can be set at an angle to the wind to carry the creature along. The clusters of different "persons" hang beneath this float. There are feeding polyps, reproductive polyps which bud off tiny medusae, and some small stinging tentacles as well as the long blue trailing ones.

As physalia sails through the water, its long tentacles stream out behind, fishing for prey. The stinging cells are among the most deadly and powerful of those in any jellyfish, and quite large fish can be paralyzed by them. Encoiling tentacles pass the prey up to the many feeding polyps which spread their mouths over the animal like a mass of digestive suckers.

Amazingly enough, there are some small fish which live among the stinging tentacles of physalia and remain unharmed. Whether they are immune to the deadly sting or manage to avoid touching the tentacles is not known, but they do gain protection from enemies while sheltering underneath the jellyfish. Many other sea jellies, including sea anemones, have clusters of small fish living among their tentacles. They too gain protection and it is thought that the fish may benefit the sea jellies by causing a disturbance in the water and thus wafting food nearer.

Although *Physalia* is a dangerous predator, it is not without its own enemies. These include the floating sea snail, *Janthina*, the sea slug *Glaucus*, and certain sunfish which will attack and eat it. Some crabs also enjoy eating the bodies of these sea jellies when they are stranded on the shore.

Another surface drifter which is made up of a colony of polyp "persons" hanging down beneath a saillike float is the beautiful little deep-blue jelly called *Velella* or "by-the-wind sailor". No more than an inch to one and a half inches across, its oblong body has a triangular sail set diagonally across it, and beneath this are a series of air-filled buoyancy chambers arranged in concentric rings. Hanging below this "raft" are the crowded "persons." A large central polyp mouth is surrounded by some smaller polyps also with mouths which are capable of producing tiny medusae to be set free into the sea. Around the outer edge of the whole animal is a fringe of stinging tentacles.

Velella feeds on microscopic plankton animals and has many enemies, including the violet-colored bubble-raft snail, *Janthina*. This creature floats at the surface by making a buoyant raft of bubbles, and it feeds on *Velella* by gradually eating the tentacles and other hanging polyps until nothing remains but the horny float and sail. These disembodied floats drift away and can often be found washed up on the shore looking like tiny plastic toys.

Comb jellies, or Ctenophores, although not classified closely with the other coelenterates, are another type of sea jelly. They have no polyp or medusoid form, but they *do* have a jellylike transparent body, usually round, ovoid or gourd-shaped, with a mouth at one end and a balancing organ at the other.

Unlike the coelenterates, which move by muscular contraction, the comb jellies have eight longitudinal rows of hairy combplates, rather like fringes of eyelashes, which beat rhythmically like paddles, propelling the animal forward through the water. As these combs catch the light they refract it, so that waves of rainbow colors can be seen sweeping along the translucent body of the animal. All comb jellies are luminescent at night and flash in the water, as do some of the other sea jellies.

Some species of comb jelly, like the sea gooseberry, have two long trailing tentacles with which to entangle their prey, but instead of nematocysts (which none of the comb jellies possess) they have special sticky cells which, when triggered, latch onto and ensnare any passing creature. Although very small, delicate and beautiful to look at, the comb jellies are voracious feeders, sometimes occurring in the plankton community and often seriously reducing the populations of young fish, crustacea and even other jellyfish.

Jennifer Coldrey

A jellyfish, stranded on the shore at low tide, is a common sight on many beaches. This one, called *Cyanea*, is one of the largest jellyfish in the world.

When alive and active, jellyfish float gracefully through the sea, with tentacles trailing. Their shape may be flat like a mushroom or deep like a bell.

In some jellyfish the tentacles are long and thin and grow only from the edge of the bell. In others, thick frilly tentacles grow from the center around the mouth and help to push food into it.

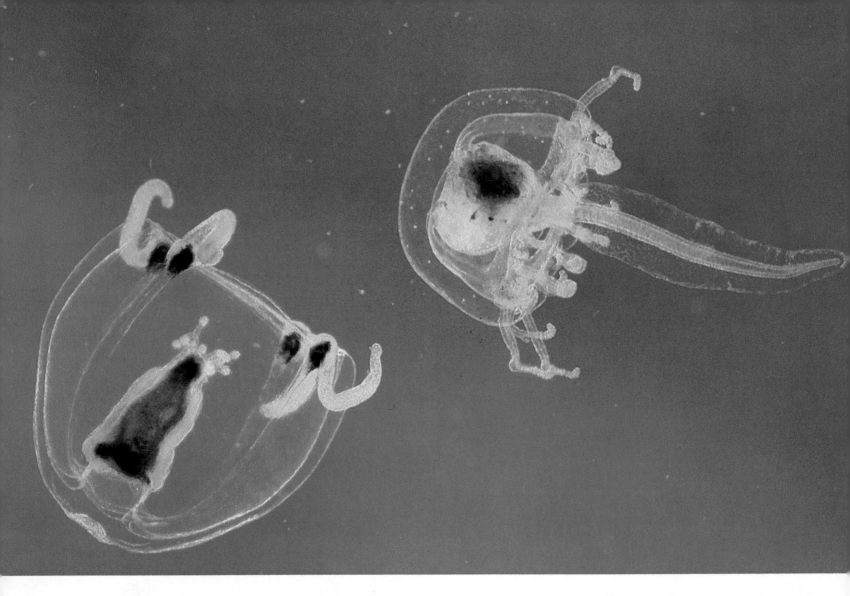

All jellyfish, however small, feed on other sea creatures. These free-floating jellies are called medusae. The one on the right, itself less than half an inch across, is eating a young fish. The food inside shows as a large brown patch in the center of the bell.

This beautiful tropical jellyfish, called *Cassiopeia*, is lying on its back on the seabed. It is a lazy feeder, and its frilly tentacles suck in food from the water like a sponge.

This tiny medusa is known as a cigar jelly because when resting on the seabed it coils up its **tentacles** and folds its body into a cigar-shaped tube.

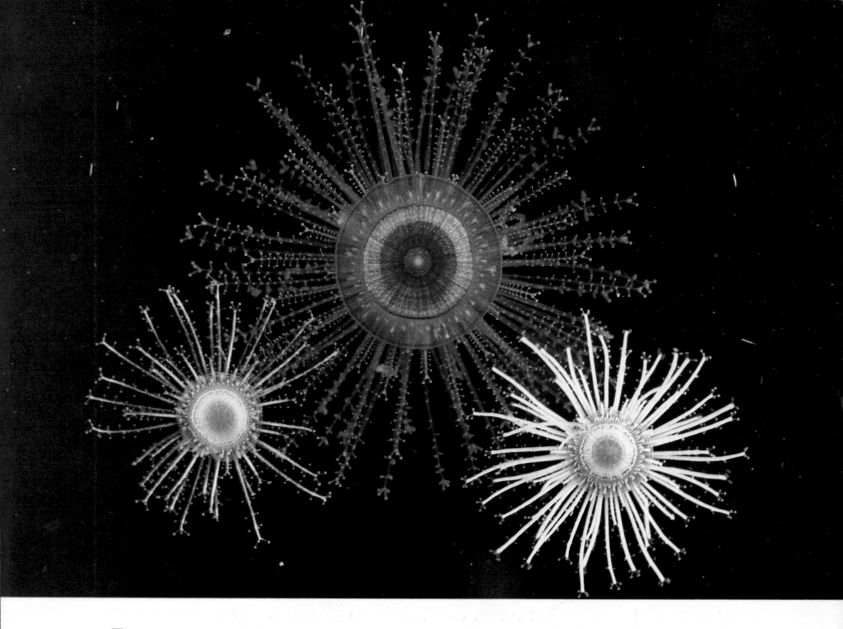

These tiny brightly colored sea jellies, called *Porpita*, float on the surface in tropical seas.

Porpita photographed from the side, with the camera just below the surface. The air chambers that keep it afloat are in the dark-blue mass.

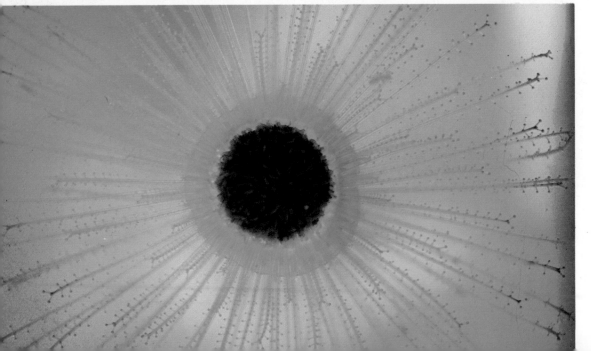

Porpita seen from below. Masses of tiny feeding polyps are visible around the central mouth. The larger stinging tentacles stretch out from the edge of the disc.

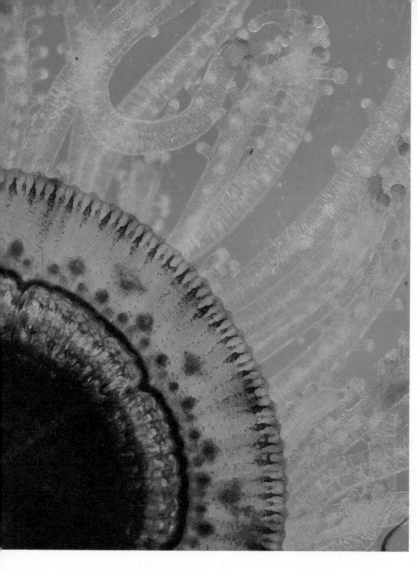

A section of *Porpita's* disc in close-up. The knoblike stinging cells can be seen along the edges of the branched tentacles.

Sea slugs are deadly enemies of *Porpita*. This one is nibbling at the tentacles.

Air chambers keep this by-the-wind sailor afloat. A triangular sail above the surface catches the wind which blows it along.

The by-the-wind sailor is so small that two of them fit easily into a thermos flask.

This hungry bubble-raft snail has attacked a by-the-wind sailor from underneath and is feeding on its tentacles.

Sometimes these animals are blown ashore and stranded on the beach. All that remains of these are the disc and the sail.

The Portuguese man-of-war, or *Physalia*, is the most feared of all sea jellies, but even it is at the mercy of the wind and the currents and can end up high and dry on the shore.

This floating Portuguese man-of-war is kept at the surface by its inflated air bladder. The trailing tentacles can sting creatures swimming as much as a hundred feet away.

Beneath the float of the Portuguese man-of-war hangs a curtain of tentacles. Among the stinging tentacles, which look like coiled springs, are others for feeding.

This close-up of a tentacle shows how the stinging cells are arranged in coiled bands.

If anything touches a cell, a stinging thread full of poison shoots out.

This fish has been trapped and is being drawn up by the tentacles toward the many mouths of *Physalia*.

Despite being one of the deadliest of sea jellies, the Portuguese man-of-war has enemies of its own, like this *Glaucus*, a sea slug which attacks and eats it.

The sea anemone is a simple form of sea jelly called a polyp. It spends its adult life attached to a rock, with **tentacles waving in the water ready to catch prey.**

Some fish can live among the stinging tentacles of sea anemones and not be harmed by them.

Corals are made by the gradual buildup of colonies of polyps. There are many different kinds of coral in the sea, some hard and some soft. This one is a soft, branching coral.

These small white flowerlike creatures are the polyp stage in the life cycle of the big moon jelly, *Aurelia*.

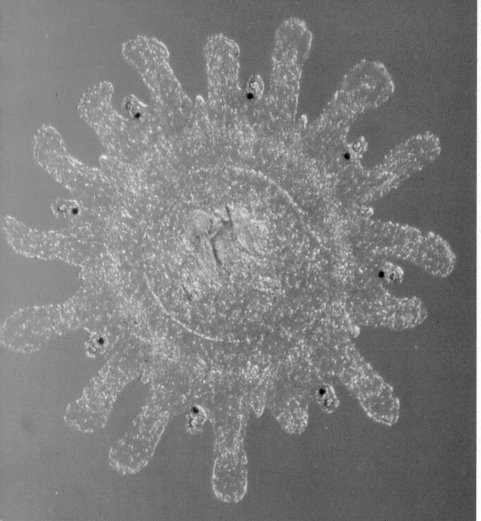

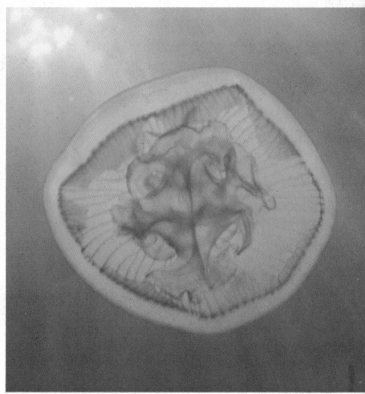

When it first floats free from the polyp the baby moon jelly is less than half an inch across. The dark spots around the edge are light-sensitive cells. The mouth can be seen in the center.

A fully grown moon jelly. Like most of the jellyfish that float near the surface, it has a pale translucent body.

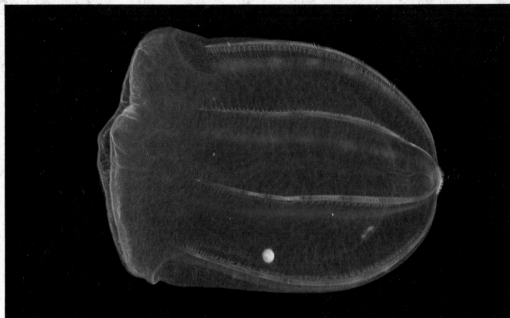

Left This comb jelly, called the sea gooseberry, has two long, branching, sticky tentacles for catching prey. On the transparent body, rows of fringed hairs, or combs, beat rhythmically, propelling the animal through the water.

Above This comb jelly, called *Beröe*, is about three inches long. It looks like a floating empty bag with an enormous mouth for catching prey. The eight rows of comb plates refract the light in rainbow colors.

Comb jellies catch and kill other sea jellies as well as young fish. This comb jelly is battling with a small medusa.

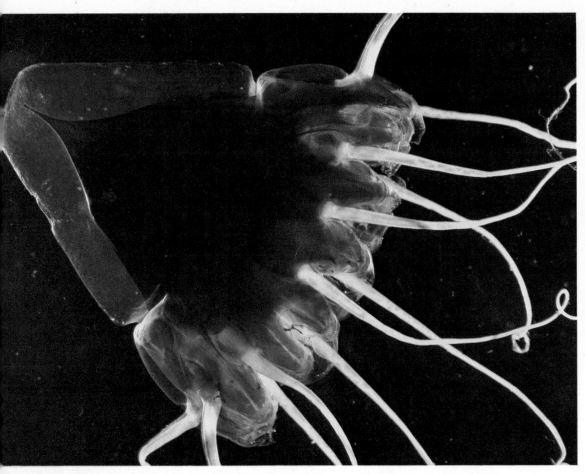

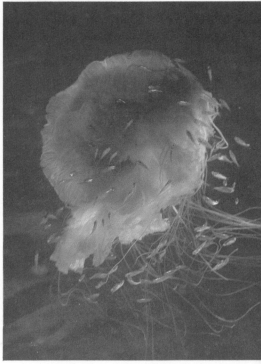

Some jellyfish live in the depths of the sea, like this beautiful dark plum-colored medusa, *Periphylla*.

Large jellyfish often have small fish swimming with them. The fish are protected from enemies here and seem to be safe from the stinging tentacles.